A Synopsis of Integrated Pest Management in Developing Countries in the Tropics

Synthesis report commissioned by the

Integrated Pest Management Working Group

Published by

NRI
Natural Resources Institute

The scientific arm of the
Overseas Development Administration

Natural Resources Institute
ISBN 0 85954 296-3

Preface

The Integrated Pest Management Task Force was set up in mid-1989 by a group of interested donors in order to assist the Technical Advisory Committee (TAC) of the Consultative Group for International Agricultural Research in its consideration of international support for plant protection. Its mission was to report on the status of IPM and make recommendations for improving its implementation and then proceed as necessary to execute those recommendations. The Task Force appointed a team of consultants to carry out a review and report the findings.

The Consultants' Report (referenced in Appendix C), which was made available to the TAC, provides a broad overview of the major issues which influence the pace of IPM adoption in the Developing World. It initiated considerable constructive debate on a number of key issues, on which agreement was not reached during the life of the Task Force.

As the work of the Task Force gradually moved into the area of implementation, as recommended in the Report, it was reconstituted as the Integrated Pest Management Working Group. Its current focus is identifying initiatives for implementation through a series of regional workshops involving key developing-country researchers, extension specialists and policy makers.

The number of donors involved in the group continues to increase, from the five original members, to twelve at the time of the most recent meeting in Paris (May 1991). The Chair and the Secretariat were originally the responsibility of the Australian Centre for International Agricultural Research. The Chair is currently held by Dr Nyle Brady, UNDP Consultant. The Secretariat has moved to the Natural Resources Institute (of the Overseas Development Administration) in the UK.

This present report has been produced for the IPM Working Group in a concise format in order to inform a wider, combined technical/non-technical audience about the IPM concept; progress and constraints. It will be used (as a workshop resource, *inter alia*) to further the implementation process. It draws substantially on the longer, more technical Consultants' Report and on the discussions of the original Task Force.

Chronology of key events:

April 1989 (Wallingford) - CABI Plant Protection Seminar.

May 1989 (Canberra) - need for IPM initiative identified and an IPM Task Force to be constituted.

August 1989 (Rome) - IPM Task Force meet, appoint team of Consultants to report.

December 1989 (Singapore) - Consultants' Report submitted to Task Force.

March 1990 (Malaysia) - first regional workshop to discuss Consultants' Report and IPM needs.

May 1990 (The Hague) - Task Force reconstituted as an IPM Working Group.

List of acronyms

CABI	Commonwealth Agricultural Bureau International
CGIAR	Consultative Group for International Agricultural Research
CILSS	Comité Permanent Inter-Etats de Lutte contra la Sécheresse dans le Sahel
CIMMYT	Centro Internacional de Mejoramiento de Maiz y Trigo
EEC	European Economic Community
FAO	Food and Agriculture Organization of the UN
IARC	International Agricultural Research Centres
IITA	International Institute for Tropical Agriculture
IRRI	International Rice Research Institute
NGO	Non-Government Organization
ODA	Overseas Development Administration
TAC	Technical Advisory Committee (of the CGIAR)
TFAP	Tropical Forestry Action Plan
USAID	United States Agency for International Development

A SYNOPSIS OF INTEGRATED PEST MANAGEMENT IN DEVELOPING COUNTRIES IN THE TROPICS

Executive Summary

(a) The Integrated Pest Management Task Force was convened to review the need for an **international co-ordinating mechanism** to expedite the implementation of IPM in the Developing World. The aim of this report is to inform a broad, technical and non-technical audience about the integrated pest management concept and the progress and prospects for its implementation in the Developing Countries.

(b) Representatives of interested donors and executive agencies (Appendix A) met in Rome. **A team of consultants was engaged to prepare working papers** which were considered by the Task Force at a meeting in Singapore and formed the basis for the present report.

(c) A **summary statement was presented to TAC** after the Singapore Meeting as a contribution to CGIAR deliberations on the crop protection functions of IARCs.

(d) It was concluded that **implementation of IPM is of primary importance in the developing countries** because of the economic and environmental benefits it brings. The Task Force was reconstituted as a Working Group in order to promote this implementation.

(e) IPM is complex and demands user participation. Much research has been poorly targeted and technology development has been inappropriate to the needs and resources of small-scale farmer clients. **IPM must be seen as a subset of crop management, itself a subset of farmer livelihood.**

(f) **Support from industrialized countries is critical in providing the tools for IPM** and in strengthening national infrastructures as a vehicle for their adaptation and integration into appropriate IPM systems.

(g) Current efforts are fragmented, overlapping and to some extent competitive. **There is inadequate flexibility within donor systems (a) for the recognition of appropriate initiatives to support and (b) for the mechanisms to respond quickly and directly.** Resources required are frequently small-scale and need delivery with high discrimination.

(h) **An international body to support and promote IPM is strongly favoured.** It should consist of senior technical representatives of interested donors (supported by a Secretariat) with the power to commit funds. The secretariat should be based upon an appropriate existing institution.

(i) The term of the Working Group was considered inadequate to provide either sufficient geographic coverage or to involve a broad enough spectrum of donors; Developing World involvement had also been less than would have been desirable. **A series of workshops was proposed to verify and disseminate the results of deliberations more widely.** As a first step, ad-hoc arrangements were made to support a workshop in Malaysia associated with the MAPPS conference (Ferrar, 1990). Further workshops for west Africa, east/central/southern Africa and Latin America are in hand.

(j) This report summarizes the considered views of the Task Force although, as there has not been an opportunity to discuss it fully, it does not imply endorsement by members or organizations which supported the Task Force.

Introduction

1. Competition between man and other organisms for the outputs of agricultural systems has long been recognized; early agriculture for the most part lacked effective means of intervention and evolved cropping systems that were diverse in space and time as a hindrance to pest build-up. Such systems were generally sustainable and of low, but adequate, productivity. This situation underwent a dramatic change as a result of increasing use of artificial fertilizers, new seed varieties and with the advent of synthetic pesticides, all of which stimulated yields on adopting farms. The new found ability for crop protection opened the door to an economic order in which agricultural technology flourished, supporting (together with health improvements) unparalleled population growth and urbanization. The agricultural revolution of the last four decades has had far-reaching impact.

> Current estimates put global losses due to pests (insects, nematodes, diseases and weeds) at US$ 300 billion annually. This equates to around 30% of potential global food, fibre and feed production. Losses each year in grain crops alone are estimated to exceed the gross and net food grain deficits of the developing world.

2. Pesticide resistance and resurgence phenomena threatened the security of agricultural production and led to a realization of the inherent instability of such a unitary approach to pest management. Intensive research was undertaken to promote the development of new means of control; manipulation of predators and pathogens, exploitation of host plant resistance, the use of behaviour modifying chemicals and a search for crop configurations less susceptible to pest attack. Sadly, these have been viewed principally as alternative approaches to pest control rather than component technologies; their success has been often dramatic but nonetheless limited.

3. Integrated Pest Management became the catch phrase of the seventies and the acronym the icon of good practice. It has remained elusive at the operational level and some analysis of its characteristics is unavoidable in the context of this report. What then is IPM and how does it differ from the practice some scientists still prefer to call pest control? Definitions abound, but the core of IPM is the development of a set of practices that maintain pest populations at a level below that which causes economically significant losses; it emphasizes minimal intervention — particularly with synthetic biocides — and husbandry of natural regulating mechanisms be they biological or cultural. This is the key to the choice of the term pest management over pest control; management implies continuation of the pest within a balanced system that itself imposes control whereas control suggests direct intervention with little concern for sustainability.

4. IPM is essentially non-prescriptive. The practitioner must 'think' and not 'do' IPM; he must therefore have at his disposal a range of appropriate technologies and adjust the mix according to his perception of the problem. For the resource-poor farmer in the developing world such perceptions are profoundly influenced by external factors; pest management is a subset of crop management, which is itself a subset of farmer livelihood. The failure to realize and allow for the socio-economic context in which a farmer must operate has been a major contributory factor in the failure of science-led research to provide workable IPM solutions.

5. It is important to note that IPM as discussed here is not a universal solution to crop protection needs. A particular, and extremely important, instance is in the control of migrant epidemic pests such as locusts and armyworms. By definition such pests cannot be 'managed' and the objective of control, most often undertaken by governments at national or regional level, must be to minimize the impact of outbreaks by whatever means of intervention is applicable; this will usually be the use of pesticides.

6. The holistic approach to IPM has created difficulties at the technical level. The bulk of IPM researchers are the product of a narrow background. They identify themselves as an entomologist or pathologist; perhaps a virologist, weed scientist or nematologist; even an ornithologist or rodent expert. It follows that each will provide a different perspective on the word 'pest'. Most will be confident that they are able to take a view on how a variety of management techniques can be integrated to provide an effective solution in the context of their specialism; few would claim to have the skills to anticipate, or even detect, negative impacts on other components of the pest complex. There are, for example, precious few rice pest experts though there are many rice entomologists and rice pathologists. It is generally accepted that, ideally, IPM seeks an approach that takes account of interactions between all biological agents that cause loss as well as an appropriate mix of control technologies. Only rarely will one pest so dominate a situation that IPM is an achievable target for workers from a single crop protection discipline.

7. It is not difficult to see why practical IPM is hard to achieve; it demands an understanding of systems and the parameters that govern their dynamics far beyond that required to kill pests. This is the area in which our knowledge is weakest, particularly in developing countries of the tropics. Progress can only be made through exhaustive — and often wearisome — research. This is expensive and demanding on skilled manpower; it is also difficult to fund in a climate where expedient short-term solutions are also necessary and provide a much more attractive proposition to the holders of the purse. In addition, and equally challenging, is the crucial requirement to translate research results into meaningful IPM strategies and to develop methods for getting these adopted by farmers. The FAO programme in rice (see para. 12 below) is an excellent example of how this can be achieved.

8. The weight of current evidence suggests however, that the technical problems are capable of solution and that these solutions can be presented in a way that can be implemented effectively by farmers. The environmental, health and economic costs of a continued reliance on pesticides rule out this option in much subsistence and smallholder agriculture in tropical Developing Countries. Pesticides will continue to serve as a cornerstone of IPM, but such a powerful tool can no longer be squandered indiscriminately in the name of increased yield.

In 1988, worldwide spending on pesticides was over US$ 20 billion: 8.9 billion on herbicides; 6.1 billion on insecticides; 4.2 billion on fungicides and 1.3 billion on other pesticides.

Pest management in the Developing World

9. The potential economic and social benefits of IPM are high even where the principal outcome is limited to a reduction in pesticide use through improved and timely application. For developing countries the gains to be derived in foreign exchange and reduction in operator hazard through eliminating *unnecessary* or *inappropriate* use of agrochemicals are very considerable.

10. Furthermore, the dynamics of tropical agro-ecosystems differ quite markedly from those of their temperate counterparts. More production is invested in standing biomass; this makes them more susceptible to perturbation and degradation. Biological processes feature more prominently in maintaining stability where climatic factors are not primary determinants of temporal cropping profiles. An illustration makes the point. Chemical control contains the brown planthopper even under intensive rice cultivation in temperate Japan and Korea. Intensification of rice in tropical SE Asia established this species as a pest because applying the temperate solution led to insecticide-induced resurgence and the creation of a 'super-pest'. A gross oversimplification, but in essence the loss of natural enemies in

temperate rice had marginal impact and the balance was in favour of the pesticide; in tropical rice this factor was of overwhelming importance.

Photo: CABI

Research at IRRI has shown that although predators, such as the spider Lycosa pseudoannulata, *played an important role in preventing pest outbreaks, they were being eliminated by pesticides.*

11. Against this background it is clear why many attempts have been made to implement IPM in developing countries. Few have had more than limited success and it is difficult to draw general conclusions as to why this is the case from experience to date, not least because failures are rarely subjected to critical appraisal. The Working Group Consultancy addressed this problem by devising a questionnaire which was issued to scientists with extensive experience of pest management in Developing Countries. A detailed analysis of these data is presented in Annex III of the Consultant's Report (see Iles, 1990); it has yielded some valuable insights and some of the general lessons and principles are reviewed here.

12. The FAO IPC programme for SE Asia under Peter Kenmore has broken new ground and provides perhaps the best example of a tropical crop of developing countries, irrigated rice, in which recent developments have closely followed the IPM concept. The programme has adopted a holistic approach

Rice is arguably the world's most important crop since it is the staple food of half the world's population. It is grown on 148 million hectares of land globally of which Asia claims the largest proportion (126 million ha) followed by South America (6.8 million ha) and Africa (5 million ha). The balance is made up by North and Central America, Europe and Oceania. Asian countries therefore contribute nearly 90% of the global total while supporting 60% of the world's population (IRRI, 1985).

to addressing crop protection issues. Recognizing the importance of policy-makers, researchers, extension workers and farmers within any system, it has successfully involved each group in the development and implementation of integrated technologies for crop protection. As a result, many farmers have been trained, appropriate technologies have been adopted, pesticide subsidies removed (in Indonesia) and pest management policies are being influenced within the SE Asia region. Although still constrained by limited availability of appropriate technology and over-emphasis on entomology, it is significant that these achievements have been possible only through the evolution of an appropriate policy environment. The case is analysed in detail by Paul Teng (see Appendix C) and has provided much needed reinforcement of the conviction that farmers are able to appreciate 'ecological principles' and both recognize and husband natural enemies in their fields.

Photo: CABI

During the 1970s Nilaparvata lugens *became a serious pest in SE Asia as a result of injudicious use of pesticides aggravated by planting of susceptible rice varieties. This photograph shows 'hopper burn' in a field of susceptible rice surrounded by plantings of IRRI resistant varieties in Java.*

13. It is clear that pesticide use, particularly against insects, overshadows all other control technologies as a means of crop protection in the Developing World. This is true even for crops such as rice, where significant progress has been made in the breeding and adoption by farmers of resistant varieties.

> Host Plant Resistance (HPR) has been the basis of plant protection for centuries. It is also the main means of technology transfer, via improved seeds, to rice farmers all over the world. IRRI genotypes are probably the most widely grown of any crop. Although much rice HPR is of the successful single gene type, frequent insect and disease outbreaks reveal its weakness when used as the sole method of plant protection.

14. Widespread adoption of the IPM concept was stimulated and necessitated by the negative effects of injudicious pesticide use. The following classification of the status of pest management situations

based on the level of pesticide intervention and resultant stress, helps to explain how this comes about.

non-use: typically rainfed subsistence agriculture in which indigenous cultural controls are presumed to feature; pest damage may or may not warrant intervention but pesticides are unavailable for cultural, political or socio-economic reasons.

balanced use: need-based application of pesticides in conjunction with other control technologies (a loose definition of IPM); alternatively, pesticide use constrained by external factors which if modified would lead to a shift into the next category.

excessive use: the beginnings of the familiar 'pesticide treadmill' characterized by increasing pesticide use to maintain yield stability in the face of developing pesticide resistance and resurgence problems.

pesticide crisis: here resistance to pesticide is well established and yet dependence on chemical intervention is complete, with no natural buffering of pest populations other than climatic variability; minor inefficiencies in pesticide application and management lead to severe crop damage. Problems may become so extreme that cultivation must be abandoned.

The stages are recognizable as an evolutionary series that often accompanies the intensification of agriculture.

15. Malcolm Iles (see Appendix C) deals with detailed points identified by questionnaire respondents. They are presented by crop. This in itself reflects our failure to take sufficiently systems-based approach to the problems of pest management, but there is much valuable information synthesized and specific weaknesses in both available technology and its application are highlighted. It is noteworthy that one of the appendices in Annex III of the Consultants' Report diverges from the commodity approach and is concerned with the parasitic weed *Striga* as a problem across commodities and thus farming systems. The importance and current neglect of weed management as an integral part of IPM is brought sharply into focus here and elsewhere.

The identification and utilization of crop varieties having at least reduced susceptibility to attack, is not complete immunity, has been the goal of many research programmes since the 1940s. Unfortunately, despite substantial research effort, *Striga* species continue to cause problems. However a success story has been the discovery of a high-level of resistance in cowpea to *S. gesneriodes*. Lines in which resistance is based on a single dominant gene are now being utilized at IITA and other national research institutions. A major challenge still remains the discovery of true resistance to *Striga* in maize. IITA has been successful in identifying potentially valuable sources of tolerance to *S. hermonthica* although for pearl millet there has been little progress.

Adapted from Parker, C, 1991. *Protection of Crops against Parasitic Weeds.*

16. The major themes that pervade the responses are presented here. The reader is referred to the Consultants' Report (Annex III) for specific crop and pest information. The views expressed by respondents to the Consultants' survey do not entirely support the claim that sufficient research has been done and what now remains is to apply it. Both basic and adaptive research still merit high priority, particularly in crop resistance strategies and pest monitoring, and the farming context in which technologies are to be used. Each situation is different; a country or cropping system must be examined

independently in order to identify the right balance of resource allocation between research and implant.

17. There is universal agreement that more research does not simply mean more of the same; research must shift its focus to broaden the discipline base and to accommodate the needs of farmers. The failure of researchers to adequately comprehend farmers' resource constraints and decision processes were widely cited as a major cause of poor adoption of IPM research recommendations.

> Conventional agricultural research tends to be scientist-led with the aim of producing technology that farmers can use. However, low adoption by farmers of IPM research output has led to drastic rethinking by scientists and socio-economists over the past decade. It is now considered extremely important that the farmer be involved in the research process in some way. Scientists must also recognize the socio-economic situation in which individual farmers operate. The most successful examples of farmers adopting research output are those which have maintained a high degree of farmer participation in all stages of the research process.

18. Considerable emphasis was given to the communication gap between farmer and researcher; the need to strengthen the role of intermediaries such as extension services and NGOs through better information management and flow, and through upgrading problem-solving skills is clear.

19. Almost all respondents felt that if the pace of IPM implementation is to be increased, more resources must be committed over a longer time-frame; there must be much improved co-ordination accompanied by functional specialization amongst both donors and research institutions.

Constraints to the implimentation of pest managment programmes

20. Implementation of effective IPM programmes demands the achievement of a complex series of interrelated goals and the establishment of mutually reinforcing feedback mechanisms to allow systems to evolve with experience. Identification and delivery of IPM systems, is not a linear process. It is also not prescriptive and must maintain an integral capacity for change in response to shifts in the pest complex, often themselves engendered by the management process.

21. The knowledge processes involved, though not necessarily in the order system definition is approached, are represented below with their corresponding activities:

Knowledge process	Activity
problem definition	farmer/researcher experience
knowledge generation	inventory/research
synthesis	research/technical services
adaptation	technical services/extension
dissemination	extension
adoption	farmer implementation

Constraints that limit IPM implementation operate across the entire technical, socio-economic and political environment in which the pest problem is experienced.

Technical constraints

22. IPM is firmly rooted in biology and ecology. It seeks to harness indigenous regulatory mechanisms for pest population control and exploit physiological and behavioural characteristics of pest species in their management. It also emphasizes the adaptation of indigenous 'well-adapted' or 'robust' cropping systems in the development of more productive agriculture. Such an approach demands a detailed understanding of the characteristics of agro-ecosystems and their dynamics in response to intervention.

23. An adequate knowledge base is generally lacking and much research has been confined to too few components of the system to predict the effects of control strategies. This is particularly relevant in the case of the farmers' situation. Farmers are nationally reluctant to adopt technology when they are unsure how it will fit within/disrupt their system. Since it will be virtually impossible for the crop protection specialist to anticipate the full effect of a given technology on an individual farmer's system, a choice, or 'menu' of technologies should be on offer from which the farmer can choose and with which further experimentation can be carried out on-farm.

The annual grass Rottboellia cochinchiensis *is spreading throughout SE Asia and the New World. It grows rapidly and produces abundant seeds. Manual control by hoeing is possible but time-consuming and costly.*

24. Increasing concern is now being expressed about our ability to characterize pest organisms. As methods of control become more specific and sophisticated it becomes more important to distinguish variability in order to ensure that the appropriate method is applied. The rationale is clear for management of host plant resistance and pesticide resistance; evidence is accumulating that the host specificity of parasitoids is much greater than previously recognized. The adaptation of research outputs from molecular biology to provide accurate, simple field diagnostics is a high priority.

25. Technical constraints limit the effectiveness of IPM component technologies. The disfavour with which pesticides have been viewed has tended to discourage research in the public sector into improved application technology, better targeting in space and time, and higher standards of pesticide management.

> Ultra low volume (ULV) application technology involves the pesticide being dissolved in volatile solvent oils then atomized by feeding them on to a rapidly spinning disk. The cloud of spray produced by this 'rotary atomization' is then allowed to drift downwind on to the crop. Wider swaths can be sprayed because oil-based droplets evaporate more slowly than water droplets. Some of the disadvantages of ULV spraying such as increased dermal toxicity, patchy spray cover and in some cases greater application costs, led to the development of very low volume (VLV) spraying which combines the advantages of wettable powder with ULV rotary atomization technology. VLV spraying has been particularly successful against cotton pests in The Gambia.

Photo: NRI

Citrus spraying in Thailand. Excessive volume and application rate leading to contamination of operators and water channels.

26. With few exceptions, the farmer still uses approaches to and methods for pesticide application that have been current for 20 years. There is almost universal recognition that pesticide use will continue and probably increase. If this is to be the case improved techniques are essential to safeguard the effectiveness of other control technologies. Low and Ultra-low Volume techniques, which could provide massive benefits in economy, safety and ease of use have not yet been well adapted for use by peasant farmers.

27. This is often related to plant architecture and some application specialists have grave doubts that much further progress is possible without, for example, crop breeding for appropriate physical characteristics.

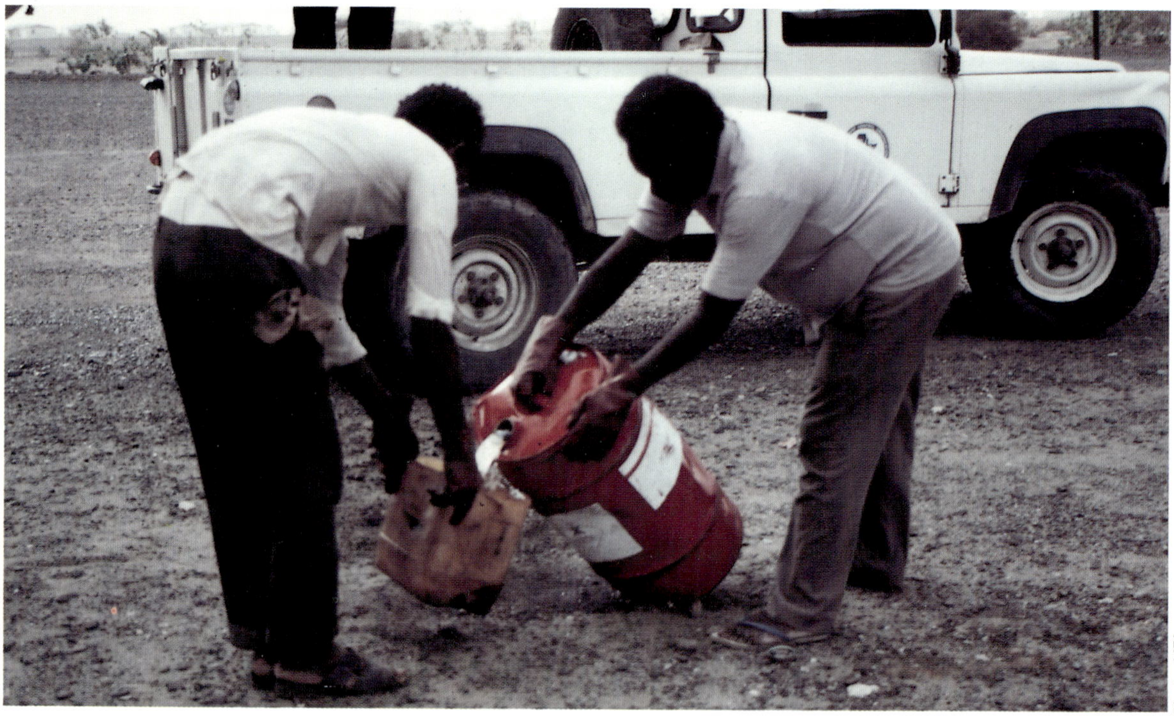

Photo: NRI

*How **not** to handle insecticide - aircraft loading using a cut-down plastic drum.*

28. Host plant resistance has made a major contribution as 'seed-borne technology' requiring minimal change in farmer practice. Even in low input farming, yields have been improved and stabilized. The major challenge is in countering the selection of pest populations able to neutralize the characters conferring resistance; this depends upon strategic deployment of resistant cultivars and the identification of polygenic resistance. An empirical approach through screening is no longer enough; the underlying mechanisms must be characterized and classified. The ability to create transgenic plants can then be fully exploited.

In Indonesia between 1979 and 1983, Host Plant Resistance (HPR) in rice (IR-36) to brown plant hopper *(Nilaparvata lugens)* was vital to suppression of the pest. It was difficult however to maintain wide resistance in varieties that are also optimal for yield quality and sale price. In 1981, two rice varieties were released with the same genes for resistance to *N. lugens* as IR-36. These varieties did not, however, carry other genetic material to confer more lasting resistance. Within two and a half years both varieties became susceptible to *N. lugens* attack.

29. Biological control is one of the fastest growing and potentially productive areas of IPM technology. The term is used here in its modern sense to cover the use of all biological or biologically derived agents. Classical biocontrol based on release of exotic or supplementation of indigenous natural enemies has had striking successes but is heavily constrained as an IPM component by problems of rearing and supply, particularly for insects. Bacteria, viruses and their toxins have presented similar difficulties compounded by problems of formulation, application, persistence and potential side effects.

Within ten years of its accidental introduction into Africa from South America, cassava mealybug was a major pest. For subsistence farmers who did not use chemicals, a biological control solution was considered the best option. Researchers found a small wasp, *Epidinocarsis lopezi,* which was not only a parasite of, but a host feeder on, cassava mealybug. Mass rearing and distribution techniques were developed at IITA; by 1987 *E. lopezi* was established in 90% of Africa's cassava-growing region. Estimates of reduction in losses range from 50% to 90%.

30. The use of behaviour-modifying chemicals has found wide application, particularly with insect pheromones, as a tool of great importance in a number of pest management systems. Pheromones may be used directly in pest control but perhaps play an even more important generic role in facilitating pest monitoring, the fundamental requirement of IPM. Here the constraint lies in identifying appropriate, cost-effective methods for field deployment and in relating catches to action thresholds.

In Egypt, control of cotton pests such as *Pectinophora* sp. by slow release pheromone formulations has resulted in increased honey production as well as an increase in boll weight. Moreover, the cost of pheromone formulations are 20% lower than insecticide.

Socio-economic constraints

31. The point has been made earlier in this report that the practice of IPM puts high demands on the user. The technologies must first be available, accessible and comprehensible; they must then be assembled to meet the particular needs of pest management problems that vary constantly. Although it is possible to identify IPM 'packages' of fairly wide applicability, true IPM more closely resembles a technology basket from which intelligent choices must be made by the farmer.

32. Farmers are the ultimate users of IPM technologies. They need to be able to make reactive decisions in order to respond to and solve a particular problem. Compared with IPM approaches, prescriptive methods of pest control based on pesticides simplify farmers' decision making. IPM demands greater skills of farmers in their being able to select (from a choice of) options which will be appropriate to their farming system *at the time the decision has to be taken.* It is clear in this context that conventional extension service activities based on prescriptive advice are unlikely to be viable. It is inconceivable that extension workers could advise under all circumstances on appropriate, *timely* action in the context of the individual's farming system.

The presence of millet amongst groundnuts reduces the risk of groundnut rosette disease.

Excessive pesticide use is characteristic of brassica production around the world because of the cosmetic nature of pest damage. Residues on harvested plants pose a direct risk to consumers.

33. Farmer education and a sense of their ownership and control over the new IPM technologies are essential for future success. Farmers manage complex cropping systems and experience has shown that they are capable of understanding the principles that must be applied to achieve IPM. The technologies themselves must be presented in a form that enables the farmer to see how they may be fitted into his or her farming system. This is a major challenge for adaptive research and demands a much improved knowledge of the farmers' decision process and the ability to produce technologies to meet the needs of a range of farmer types.

In Malaysia, vegetable farmers benefitted from discussions and demonstration trails on control of *Plutella xylostella* and are now aware of the usefulness of parasitoids against the moth. Many held back spraying of their cabbage crops until a higher level of *P. xylostella* is observed resulting in reduced insecticide use.

Institutional constraints

34. IPM is inherently an inter-disciplinary, multifunctional approach to solving pest problems. Current institutional structures in both developed and developing countries do little to simplify the task of the farmer practitioner. Components of the problem, in both disciplinary and operational terms, are commonly abstracted to form the principal axis for the organization of public sector institutions. The management of research, extension and technical support services are frequently operated independently of one another, centred in different institutions and often with conflicting goals and interests. These activities are almost always under-resourced and unable to compete with the commercial sector.

A joint CILSS/FAO/USAID project looking at control of *Striga* in West African countries noted institutional constraints which manifest in poor co-ordination and co-operation between plant protection services, research institutions and extension services. All three organizations were located in different ministries and appeared to be more interested in their own independence and domains than in co-operating with each other. This resulted in the setting of partisan priorities, poor use of available human and material resources, waste of technical assistance funds and ultimately, no IPM at farmer level.

Political constraints

35. An appropriate policy environment is essential to the establishment of IPM. The need to generate self-sufficiency in food production is a powerful imperative for most, if not all, developing countries and there remains a strong tendency to support simple short-term approaches with a rapid pay-off such as those dependent on pesticides. The situation is further complicated by the vested interests that surround the pesticide industry and its undoubted political and commercial influence in developing countries. Pesticide subsidies have done a great deal to undermine their rational and judicious use, eventually to the producers' disadvantage.

Governments of Developing Countries subsidize pesticide production and sales in a variety of ways: through access to foreign exchange on favourable terms, through tax exemptions, through easy credit and below-cost sales by government-controlled distributors. The median level of subsidy is around 44% of retail cost and in large countries with generous subsidies such as Egypt (estimated at 83% of full retail cost) the total cost is US$207 million annually. In Indonesia (estimated 82%) until the subsidy system was dismantled and IPM adopted as national policy, the total annual cost was US$128 million.

36. The pesticide industry is beginning to take a longer-term view and can be expected to contribute to the development of IPM in developing countries, given the reality that pesticides will continue to form a part of the solution to most problems and that agrochemical companies are now diversifying their interests into alternative approaches to pest management, particularly those based on biotechnology. A number of developing countries have declared IPM as a national plant protection policy. Such a step has major impact in facilitating change at the grass-roots level.

Criteria for successful pest management programmes

37. The definition of IPM has been the subject of major discussion during the deliberations of the Working Group and the consultancy it established. Some of the issues are considered in earlier sections of the report. Several members subscribed to the view that the term was self-defeating in its assumption that the best solution always requires the 'integration' of components. This is clearly not the case. Nor do circumstances always permit the adoption of a pest management strategy that could be called integrated - or even optimal; such a solution may nevertheless represent a great advance over no action and be therefore desirable. The term 'Appropriate Pest Management' was coined to allow for such situations. Serious consideration was given to recommending its adoption but the majority opinion was that the disadvantage of introducing a new term outweighed the merits of the reformulation. The issue is raised here because it has a bearing on the selection of criteria.

A new method of fungicide application has been developed for reducing losses from black pod and other fungus diseases of cocoa. Copper fungicide is incorporated in a water permeable collar and attached to the top of the pod-bearing trunk. Rainwater distributes copper to all parts of the tree below the collar as well as the soil around the base. The method has the advantage over conventional spraying because it releases and distributes fungicide during the rain when maximum protection is needed.

38. The success of pest management programmes may be judged against a number of individual criteria. Some of these are not exclusive to IPM, though success is more likely to be sustainable if the solutions to pest management problems accord with the philosophy that underlies the IPM approach.

39. The desired outcome of pest management is a reduction in losses arising from pest attack; this must normally be subject to the qualification that the social, environmental and economic cost of inputs must not exceed the value of the increased returns so that there is no net negative impact on farmer livelihood, the consumer or production economics. Specific criteria are associated with technology developments where:

The practice should be robust i.e. sustainable and capable of tailoring to the needs of individual farmers. It should not be dependent on a single component technology and should maximize the involvement of the farmer in decision making;

It should be responsive to the pest situation, linked through monitoring procedures and triggered by clearly identified action thresholds or indicators;

The practice should take account of interactions between different components of the pest complex; action against a particular target pest should be taken after explicitly addressing the likely outcome and implications for other pest management actions;

The component technologies must be within the technical competence and economic means of the farmer to adopt and within his capability to conceptualize;

Priority should be accorded to practices that minimize the application of pesticides and promote their safe and efficient use; explicit consideration should be given to the use of application technologies that do not have a negative impact on natural regulatory agents within the system.

The role of international development agencies in the implementation of pest management programmes

40. The international development community has shown increasing interest in promoting an IPM approach to crop protection problems; this has been at least partly induced by mounting pressures to adopt environmentally sensitive aid policies which are sustainable. Pesticides have been of major concern to environmentalists and IPM is commonly accepted as a satisfactory way of keeping their use in check. FAO has been active in this field since the 1960s and the World Bank has recently adopted IPM as the central strategy for projects involving crop protection. The climate for IPM is clearly favourable.

41. Bilateral development agencies commonly have both a funding and executive function; in some cases the latter is primarily resourced through in-house expertise, in others a national or international pool of manpower provides the necessary resource for technical co-operation. In either case there is characteristically a split between the two functions. Where funding is provided for research there are few explicit links with implementation programmes. This is a major constraint to the effectiveness of aid in science-based activities such as IPM.

42. Bilateral agencies also provide funding for multilateral organizations, which may themselves act as donors or as executive agencies. The distribution of funding, geographically and by sector, and between bilateral and multilateral activities is determined predominantly by political considerations.

43. These factors combine to create a weakly articulated and heavily bureaucratized framework that militates against a holistic approach such as IPM. There is a clear need to co-ordinate the efforts of agencies concerned to promote IPM and capitalize on their collective skills and experience. This was the rationale for the establishment of the present Working Group and a tentative mechanism for achieving this objective is set out in a later section of the Consultants' Report.

44. Technical co-operation must be the principal route through which development agencies can promote IPM. IPM is knowledge based and the generation of descriptive models, information management systems and component technologies exerts high intellectual, technical and technological demands. The scientific base needed to service this fully within the time scale in which it will be required will be beyond the resources of most developing countries. The initial effort must therefore come from the developed

countries in close collaboration with counterpart scientists from developing countries.

45. IPM is also a dynamic concept. Strategies, technologies and their performance must be constantly monitored, reappraised and revised; IPM cannot be delivered as a package. The technical capacity to perform this function must be built within the countries that operate the system. A high investment in training with emphasis on in-country practical experience based on relevant problems is required.

46. Technology generation is primarily a research activity and is inevitably science led; it must, however, be responsive to real pest management needs and not simply represent a vehicle for elegant applications of research findings. Perhaps the critical role for development agencies in facilitating IPM is to assist in the articulation of these needs through helping to characterize the agronomic, social and economic context in which pest problems are expressed.

47. This calls for a broader multidisciplinary framework within the agencies themselves and, in particular, a greater role for agronomists, sociologists and economists in defining the specifications against which control technologies are to be developed. This grouping of skills is also a prerequisite for adaptive research leading to workable strategies for IPM. The starting point for IPM must always be existing practice and development agencies must be more prepared to invest in preparatory studies leading to adequate problem identification.

An international framework to support and promote IPM

48. There are many interested groups that contribute to the formulation and delivery mechanism required to implement IPM. Their roles and relationships are reviewed by Clive James (see Appendix C). For the developing countries these included: farmers and farmer groups; national extension services; crop protection technical services; national R & D institutions; national governments and NGOs. In the industrial countries and the international arena the following were considered: international agricultural research centres (including CGIAR institutions) non-associated centres and others; regional institutes and organizations; multilateral and bilateral aid organizations; industrial country R & D organizations and the private sector.

49. Like the technical pursuit of IPM goals, the institutional framework on which their implementation depends suffers from fragmentation rather than integration of effort. Lines of communication are poorly developed and activities are compartmentalized both vertically and horizontally. Thus duplication of effort occurs and appropriate feedback mechanisms between, for example, research, extension and implementation do not exist.

50. The Working Group considered a number of options identified by the James report and reached broad agreement that there is no case for establishing new institutions; rather, a framework to facilitate the interactions between existing institutions needs to be put in place to maximize effective collaboration. This would probably take the form of a small authoritative group of professionals representing the major donors with interest and expertise in pest management.

51. Such a body could be based on the existing Working Group, appropriately modified to broaden donor and technical representation. The objective would be to generate an IPM Action Plan analogous to the Tropical Forestry Action Plan (TFAP) which seeks to achieve this for tropical forestry and to carry this forward to the definition and implementation of these actions. The immediate objectives would include promoting collaboration between new and existing projects funded by individual donors, and the provision of supplementary resources in support of national programmes. In the longer term joint

programmes would be designed exploiting the comparative advantage of collaborating bodies and based on problem identification in representative farming systems.

52. The revised Working Group should be based on an existing organization. The most appropriate organisation to serve as a base for the Working Group Secretariat is still under discussion. It would provide the Group with international links, expertise in information management and the provision of support facilities for pest management. The secretariat would then provide the hub for a series of commodity/cropping system and IPM component technology-based networks.

53. Such an initiative would include collaboration with the private sector and IARCs in maintaining catalogues of sources of host plant resistance, monitoring of resistance breakdown, development of insecticide resistance, biological control agents and pest status. It would also include the provision of an international source of reference for pest identification and the characterization of variability.

APPENDIX A

Task Force members

Australia

Dr George Rothschild (Chairman)
Dr Gabrielle Persley (Secretary)
Australian Centre for International Agricultural Research (ACIAR)
GPO Box 1571
Canberra, Australia

Canada

Dr Greg Spendjian
International Development Research Centre (IDRC)
PO Box 8500
Ottawa, Canada

FAO

Dr Lukas Brader
Food and Agriculture Organization of the United Nations (FAO)
Plant Production and Protection Division
Via delle Terme di Caracalla
00100 Rome, Italy

United Kingdom

Dr Denis Greenland
CAB-International
Wallingford
Oxon OX10 8DE
U.K.

Mr John Perfect
Natural Resources Institute (NRI)
Chatham
Kent ME4 4TB, UK

United States of America

Mr William Furtick
United States Agency for International Development (USAID)
State Department
Washington D.C. 20523
USA

APPENDIX B

Consultants

Professor Michael Way (Team Leader)
Imperial College at Silwood Park
Ascot, Berkshire SL5 7PY, UK
(Phone 44-0344-23911, Fax 44-0344-20094)

Dr David Groenfeldt
The Conservation Foundation
1250 24th St. NW,
Washington, DC 20037, USA
(Phone 1-301-565-5936, Fax 1-202-293-9211)

Mr Malcolm Iles
NRI, Central Avenue, Chatham
Kent ME4 4TB, UK
(Phone 44-634-883 054, Fax 44-634-880 066)

Dr Clive James
c/o CIMMYT
Lisboa 27, Apdo Postal 6-641
06600 Mexico 6, DF, Mexico
(Phone 52-595-45499, Fax 52-595-41069)

Professor Paul Teng
Department of Plant Pathology, College of Tropical Agriculture
University of Hawaii, 3190 Maile Way
Honolulu, Hawaii HI 96822, USA
(Phone 1-808-948-7759, Fax 1-808-948-8054)

Mr John Terry
Weed Research Division
Long Ashton Research Station
Bristol B51 898 F
UK
(Phone 44-272-392181, Fax 44-272-394007)

APPENDIX C

References

Way M.J., Groenfeldt, D., Iles, M.J., James, C., Teng, P.S. and P.J. Terry. (in press). *Integrated Pest Management in Third World Agriculture: experience and prospects.* A report submitted to the International Pest Management Task Force.

Teng, P.S. (1990) *Integrated Pest Management in rice - an analysis of the status quo with recommendations for action.* (Annex I)

James, C. (1989) *Implementation of Integrated Pest Management (IPM) in the Third World - institutional aspects.* (Annex II)

Iles, M.J. (1990) *Analysis of IPM survey responses.* (Annex III)

Ferrar, P. (1990) *Report on South East Asian Regional IPM meeting. Genting Hotel, Malaysia:* 24 March 1990.

Copies of these documents can be obtained from:

Malcolm Iles
Secretary, IMP Working Group
Natural Resources Institute
Central Avenue
Chatham Maritime
Kent ME4 4TB UK

Phone: (44) 634 883054 Fax:
(44) 634 880066/77